Concise Guide to Climate Change

Tobias Moore

Copyright © 2019 Tobias Moore
All rights reserved.
ISBN: 9781713221067
sohmpublishing.com

DEDICATION

To our future

My other books

A Different Road: From Bum to Mystic

Guide for the Homeless: Skills for Surviving the Streets

Concise Guide to Companion Planting

Concise Guide to Type II Diabetes (Spring 2020)

Concise Guide to Climate Change Adaptation (Summer 2020)

CONTENTS

Global Warming	7
Extreme Events	21
Ocean	33
Fire and Ice	41
Misconceptions	47
Things we can do	53
Graph Sources	75
Resources	77
About the Author	83

Climate Change is a real issue. It is not a hoax or conspiracy as some would have you believe.

Throughout this guide are graphs and images to help bring home the science. All data comes from scientific organizations and governmental agencies. It is up to you to take this information and imagine for yourself where things are going.

This is the first of two guides on Climate Change. This guide focuses on the facts of Climate Change and offers hundreds of actions we can do to mitigate future emissions. The second guide will focus on the psychological barriers and traumas connected with Climate Change and offer hundreds of actions we can do to adapt to our changing world.

Side Note: I have two versions of the *Concise Guide to Climate Change*: Kindle and print. I highly suggest you buy the Kindle version if you are not giving a book away or keeping a hard copy for yourself: this will save you money and you will be doing something to address Climate Change and pollution.

Section I
GLOBAL WARMING

Deniers and Liars

American Petroleum Institute
Western Fuels Association
Global Climate **Coal**ition
Fringe scientists
Koch brothers
ExxonMobil
Chevron
Shell
BP

The Concerned

Department of Agriculture
Department of Commerce
Department of Defense
Department of Energy
Department of Health and Human Services
Department of the Interior
Department of State
Department of Transportation
EPA
NASA
NOAA
National Science Foundation
At least 97% of scientist
Pope
Catholic Climate Covenant
Unitarian Universalist Ministry for Earth
Evangelical Environmental Network
Jewish Climate Action
Presbyterians for Earth Care
Islamic Climate Change Declaration
Buddhist Tzu Chi Foundation
Humane Society International
Smithsonian Institution
U.S Agency for International Development
IPCC
United Nations
World Meteorological Organization
Greenpeace
Al Gore
Sierra Club
Audubon Society
National Geographic
National Wildlife Federation
Rainforest Action Network
Environmental Defense Fund

More Concerned

The Paulson Institute
Conservatives for Responsible Stewardship
Frank Luntz
The Pew Charitable Trust
350.org
C40
Greta Thunberg
Xiuhtezcatl Martinez
Many Powerful Youth Organizations
The Rockefeller Foundation
Climate Investment Fund
World Wide Fund for Nature
Friends of the Earth International
International Union of Conservation of Nature
Rhiana Gunn-Wright
Hindou Oumarou Ibrahim
Changua Wu
Forest Stewardship Council
Theodore Roosevelt Conservation Partnership
Climate Action Network
Marina Silva
Alexandria Ocasio-Cortez
Terry Tamminen
Operation Free (coalition of veterans)
American Fisheries Society
World Wild Life Fund
Rodale Institute
Todd Tanner
Jane Fonda
Leonardo DiCaprio
Don Cheadle
Gretchen Bleiler
Idle No More
Union of Concerned Scientists
Yale Project on Climate Change Communication

Even More Concerned

Robert Redford
Emma Thompson
Mark Ruffalo
Martin Sheen
Arnold Schwarzenegger
Jaden Smith
Grimes
Radiohead
Steven Rinella
Center for Food Safety
EcoAgriculture Partners
Elon Musk
Amory Lovins
David Attenbourough
Bill Nye the Science Guy
International Food Policy Research Institute
American Medical Association
American Academy of Pediatrics
National Association of Realtors
Willie Nelson
KT Tunstall
Phish
Green Day
Jack Johnson

and many, many, many more...

List of over 200 scientific organizations from around the world
http://www.opr.ca.gov/facts/list-of-scientific-organizations.html

What is global warming?

Global Warming is an observed trend of the overall temperature of the planet getting warmer. This should not be confused with the weather, which can vary greatly from year-to-year and from place-to-place. When you think about Global Warming, imagine taking temperature readings from a bunch of different places around the world throughout the year and average them out. Now continue to take those readings every year and plot them on a graph. This is what you will see.

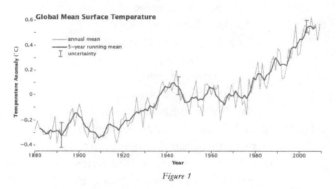

Figure 1

Notice the drastic rise in temperature over the last hundred years. What has changed in that time? Our extensive consumption of fossil fuels.

Carbon Dioxide

While there's evidence humans have used coal as far back as 1,000 BCE, it wasn't until the industrial revolution starting in the mid 1700s that coal became a primary source of energy. In the late 1800s there was a huge increase of oil and natural gas consumption starting with the first oil well in 1859 followed by the invention of the automobile.

From that point on our consumption and dependence on oil and natural gas has grown exponentially. It's almost impossible today to point to anything that is not in one way or another dependent on fossil fuels.

Why does that matter? Burning fossil fuels releases carbon dioxide (CO_2) into the atmosphere. As CO_2 collects in the atmosphere it re-emits some of the heat that would have otherwise escaped the earth's atmosphere back towards our planet. This greenhouse effect directly contributes to Global Warming.

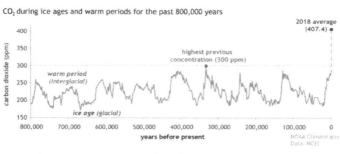

Figure 2

This chart shows the CO_2 levels for the past 800,000 years.

Notice how the CO_2 levels rise and fall over spans of thousands of years. That is the natural CO_2 cycle of the planet. Also notice in all that time the CO_2 levels never rose above (300 PPM). PPM stands for parts-per-million. Think of a million ping pong balls and 300 of those balls are CO_2 molecules. Our current CO_2 level is somewhere around 408 PPM.

Our atmosphere is primarily nitrogen and oxygen: 99% in fact. The other 1% is argon, carbon dioxide, nitrous oxide, neon, helium, methane, krypton, hydrogen, and other elements.

While 1% is not a lot, just imagine one teaspoon of salt in a cake recipe: it is perfect. If you keep adding a little bit of salt and a little bit more though, eventually it will ruin the cake.

While CO_2, methane, and other gases do not make up a large percentage of our atmosphere, if we keep emitting them, things will continue to worsen.

How do we really know about CO_2 levels? Ice core samples.

By pulling ice core samples thousands of feet deep we can measure tiny gas bubbles locked in the ice. Depending on how far down the sample is will determine how old the trapped gas is.

Because CO_2 spreads evenly throughout the atmosphere we are able to get a good idea of how much CO_2 was in the atmosphere during those points in time.

One thing I encourage you to notice throughout the following pages is the trends between all these different things and the rise of global temperature. While there is not always a one-to-one correlation, when you step back and look at the bigger picture it is hard to deny that there is some connection.

Some things to help reduce CO_2 levels
- move away from fossil fuels
- use alternative energy
- move towards all electric vehicles, tools, and so on
- regulations to drastically reduce emissions
- funding for large scale LED bulb exchange programs
- incentivizing alternative energy technologies
- making supply chains more efficient
- stronger conservation laws and enforcing them
- afforestation and reforestation projects

- encouraging old growth forest programs
- cap-and-trade
- carbon taxing
- researching ways to reduce aviation pollution through more efficient planes, different types of fuels, and so on
- all new construction focuses on efficiency
- national water conservation
- sustainable lumber practices
- buying locally
- funding organizations that are doing something positive
- connect with other concerned people
- encouraging politicians to act
- more ideas starting on page 56

Here's a list of what different U.S Agencies are doing:
https://www.c2es.org/site/assets/uploads/2012/02/climate-change-adaptation-what-federal-agencies-are-doing.pdf

Methane

Molecule-to-molecule methane is approximately 28-34 times more potent a greenhouse gas than CO_2. Like the CO_2 rise, methane is rising as well.

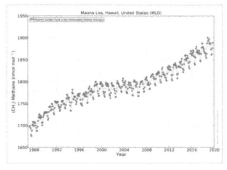

Figure 3

Source of methane
- wetlands
- wet-paddy rice farming
- livestock farming
- manure management
- biomass burning
- coal mines
- garbage dumps
- termites
- waste treatment plants
- pipeline leaks

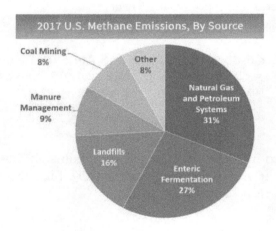

Figure 4

A few things to help reduce methane are
- capturing and converting methane into useable energy
- making sure there are not supply chain leaks
- upgrading industrial equipment
- better manure management programs
- researching better livestock diets and feed additives
- reuse and recycle
- less consumption and waste

- eat more vegetables and reduce red meat consumption
- support organic farming practices
- shifting rice cultivation practices
- more ideas starting on page 56

Nitrous Oxide

The EPA states nitrous oxide (N_2O) accounts for 5.6 percent of US greenhouse gas emissions. Nitrous oxide is somewhere in-between 265-298 times more potent a greenhouse gas than carbon dioxide. On top of that, it depletes the ozone layer, that protective little bubble around the earth that shields us from some solar radiation.

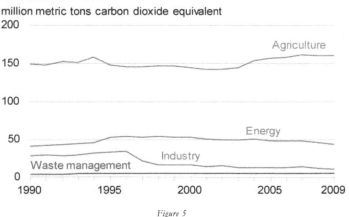

Figure 5

Where it comes from and is used for
- medicine and dentistry
- motor sports
- deep sea diving
- released naturally from soil and water
- agriculture fertilizers

17

- manufacturing of some chemicals and nylon
- power stations
- manure management
- transportation
- melting permafrost

Things we can do to reduce nitrous oxide emissions are
- better farming practices (less tilling, improving fertilization techniques, use only what's needed, manage runoff, using organic rather than synthetic fertilizers)
- reducing fertilizer dependence by using legumes to bind nitrogen
- better manure management practices
- reducing fuel usage
- electric cars
- commercial catalytic systems to decompose N_2O to N_2 and O_2 while using heat in the conversion to produce steam power
- pasture cropping
- more ideas starting on page 56

Fluorinated Gases

Unlike the other gases, fluorinated gases have no natural source: it's all us. The EPA states fluorinated gases can be thousands to tens of thousands of times more potent a greenhouse gas than carbon dioxide.

Figure 6

Some major sources of fluorinated gases
- refrigerants (refrigerators and air conditioners)
- aerosol propellants
- foam blowing agents
- solvents
- fire retardants
- magnesium and aluminum production
- electronic manufacturing
- insulating gas in electrical transmission equipment

What we can do to reduce emissions
- better handling
- fixing system leaks
- substitute with gases that have less impact on environment
- research better refrigerant technologies
- encourage fluorinated gas recycling and better destruction processes
- improving aerosol technologies
- reduce air conditioning use
- better home insulation to reduce need for air conditioning
- more ideas starting on page 56

Water Vapor

Water vapor is a classic example of a feedback loop. Things warm up, more water evaporates, water vapor collects in the atmosphere, the vapor absorbs more solar radiation, which in turn causes more warming.

Just as with the other greenhouse gases, we can see from the following chart that water vapor is increasing as well.

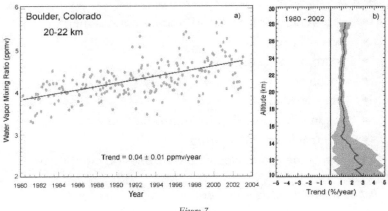

Figure 7

One tool scientists use to measure water vapor in the atmosphere is a weather balloon carrying a frost-point hygrometer. As seen from the above graph, every year has seen an increase.

Section II
EXTREME EVENTS

Section I focused on Global Warming and the greenhouse gases directly responsible for the rising global temperature. This section focuses on how the world climate is changing because of Global Warming and the human cost that comes with it.

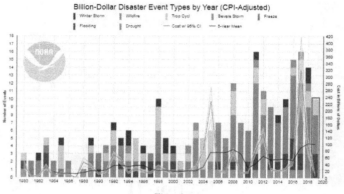

Figure 8

Beyond the cost of life, the monetary cost from extreme weather events is skyrocketing. Even insurance companies are taking note and making it harder for homeowners in certain areas to get insured (flood and high-risk fire areas). From 2015-2018 over 340,000 homeowners in California alone had their insurance dropped because they lived in high risk fire zones

In three years (2016-2018) the total cost from climate disasters in the United States was over $450 billion dollars, and billion+ dollar disasters are getting more and more common.

As of this writing there has been three historical floods in Venice in the past two weeks and up to this point (11/2019) in the United States there has been 10 billion+ dollar weather disaster events (https://www.ncdc.noaa.gov/billions/). The record breaking Midwestern floods alone is estimated to cost nearly 3 billion dollars, and that doesn't account for the human suffering.

Floods

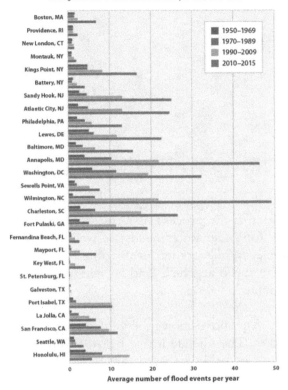

Figure 9

This chart is sobering. Every city on the list has seen an increased number of floods. In some cases the numbers have doubled from decade-to-decade.

According to National Oceanic and Atmospheric Administration (NOAA), 2019 has been the wettest year on record. The Midwest flood affected over 14 million people.

Desertification

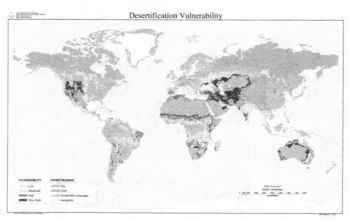

Figure 10

Desertification is the process of fertile land losing its ability to sustain our food and water needs, as well as making the land uninhabitable. While some forms of desertification happen naturally, the speed at which land is changing into desert today is unprecedented.

The USDA map above shows the vulnerable areas in the U.S to desertification. The satellite image below of Ningxia, China is a good example of land destroyed through overuse. The need to protect the land and waterways from the devastating effects of desert encroachment is of high priority. Once land has succumb to desertification, it is extremely difficult and costly to reverse.

Figure 11

Contributing factors to desertification
- urbanization
- mining
- tilling and irrigation practices
- overgrazing
- depleted groundwater from farming and livestock
- deforestation
- shifting weather patterns
- higher yearly temperatures due to Global Warming
- fires
- droughts
- soil erosion
- loss of micro-organisms due to poor farming practices

Tornadoes

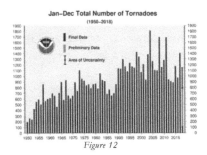
Figure 12

Notice the increased number of tornadoes after the middle of the chart. Between 1950 and 1990 there were only two years with more than a 1000 recorded tornadoes. Nearly every year since 1990 has had more.

As of November 2019, there has been 1,121 confirmed tornadoes and 39 fatalities. According to NOAA, in the month of May alone the monetary costs from tornadoes and their related storm fronts were over $3.2 billion ($3,200,000,000).

Droughts

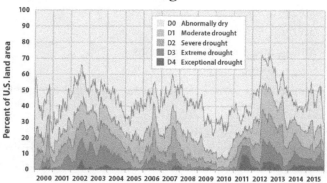

Figure 13

Droughts are devastating in many ways. Lack of water, dying crops and livestock, and increased fire risk are the most obvious consequences. Loss of top soil, dust storms, health problems, malnutrition, human deaths, human and wild animal migrations, diseases, war, massive plant loss, and sometimes mudslides when the rains do come are other consequences. Today (11/24) 60 people were killed in Kenya from a landslide.

Forest Dieback

Figure 14

While some forests have thrived with the extra CO_2 and N_2O, Boreal and Amazon forests have been dying back. This in turn reduces carbon sequestering and lessens the cooling effects forest evaporation has in the region.

Some Contributing factors of dieback
- acid rains
- droughts
- pathogens
- parasites
- heat waves
- fires
- Global Warming

Hurricanes
Atlantic Basin Storm Count
(Including Subtropical Cyclones)

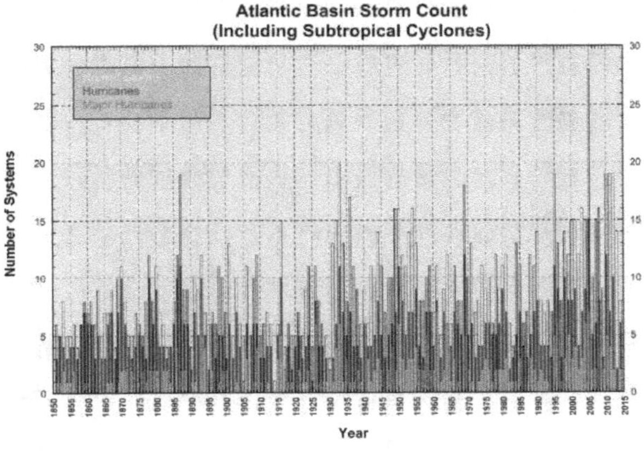

Figure 15

Notice the 15 number mark in the middle of the chart. Starting in the 2000s we see an increase in the number of events as well as the intensity of storms. In the 150 years from 1850 to 2000 there are seven years where the U.S. experienced more than 15 major storm events. Since 2000 six years had over 15 major storm events with 2015 breaking records with the most major storms in a year ever recorded.

Global deaths from extreme weather events in 2018

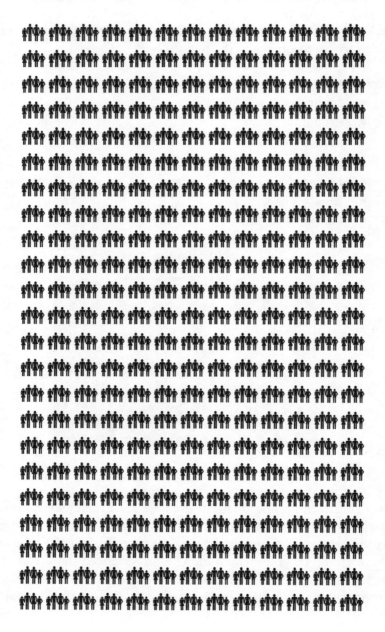

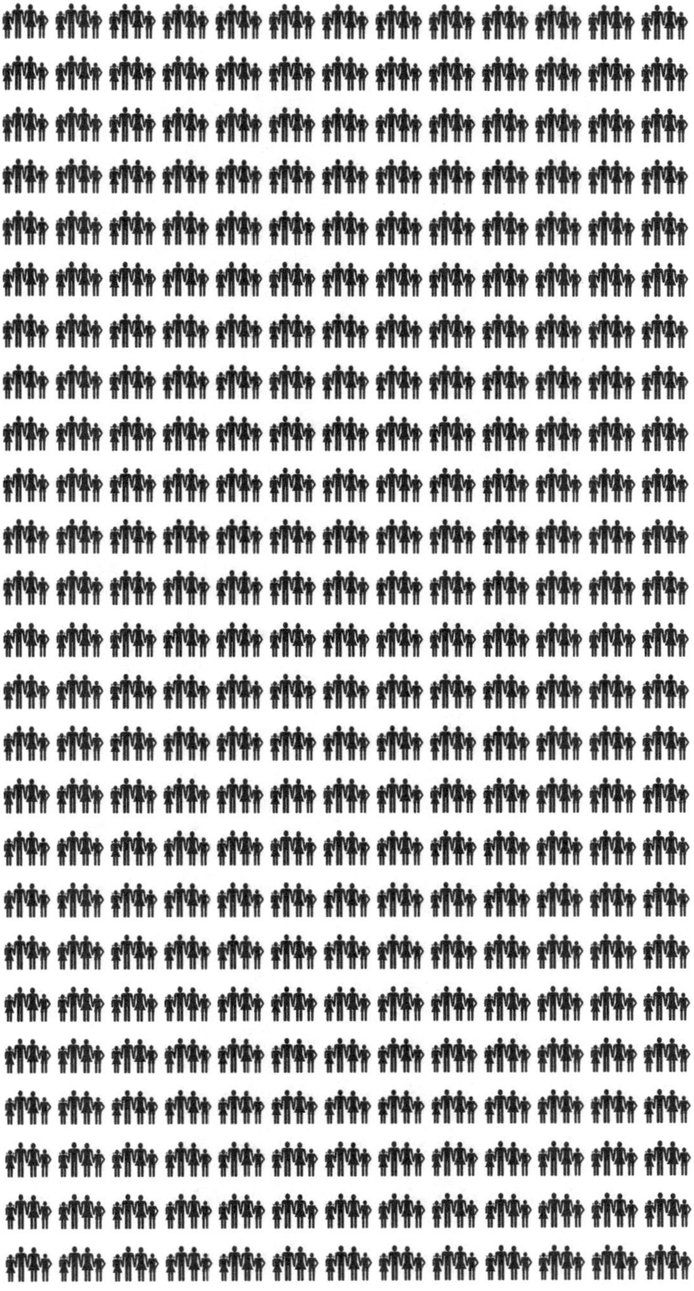

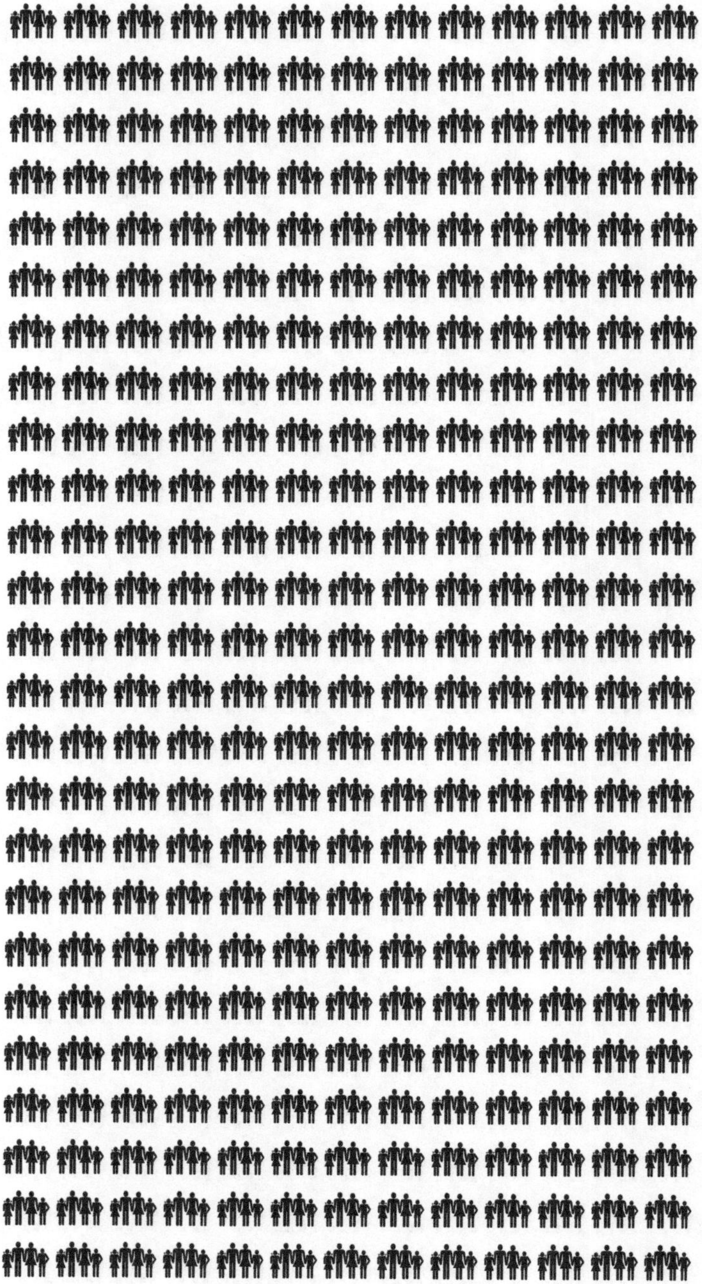

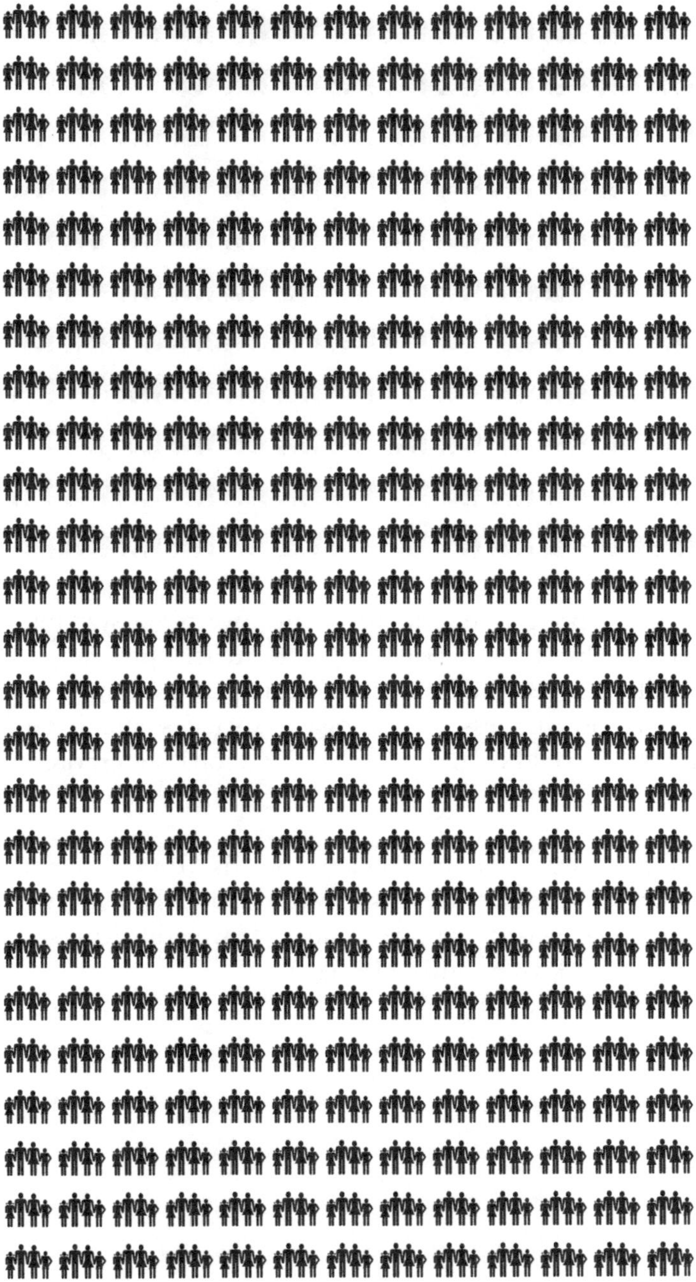

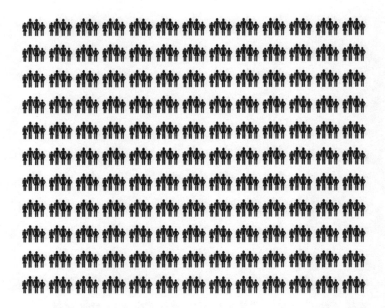

Thankfully, predictive science, better warning systems, and improved disaster preparation and relief programs have drastically reduced the number of deaths from servere weather events. Although, less technological societies still have high numbers. It would be nice to see rich nations help more in this area.

The above deaths do not include deaths from droughts, air pollution, or diseases like Zika, Malaria, and Dengue that are migrating because of Climate Change. If we added those in we would need at least another 115 pages.

Section III
OCEAN

Not only are we seeing an increase in severe weather events as pointed out in section II, we are also seeing major changes in the ocean. The following graphs highlight some scary trends we are currently witnessing with the ocean. These include getting hotter, rising sea levels, becoming more acidic, dying coral and phytoplankton, majors shifts in migration patterns, oxygen dead zones, slowing and changing currents, and massive increases in extinction rates.

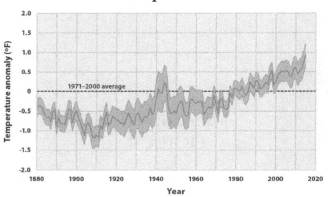

Figure 16

Some effects of rising ocean temperatures
- as water warms it expands and causes the ocean levels to rise
- more flooding
- more precipitation which gives rise to more water vapor which in turn contributes to Global Warming
- more extreme weather events
- shifting migration patterns effecting human food supplies
- killing off coral reefs
- more sea ice melting

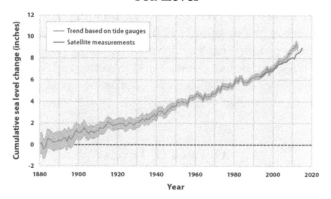

Figure 17

Major contributing factors of sea level rise
- water expands as it warms
- melting glaciers
- melting Greenland and Antarctic's ice sheets
- coastal erosion

The ocean continues to rise about .13 inches a year. That might not seem like a lot, but just imagine a child playing and making waves in the tub and every minute the water keeps rising .13 inches. It's not long before that water starts splashing on the floor. This is what more and more intense storms are doing to our coastal cities. Storm drains back up because the water cannot leave the city, which in turn causes major flooding.

That's not all though, the tub keeps filling up until it eventually spills over. When the water comes on land it causes erosion, land sinkage, flooding, contamination of water supplies, destruction of property, and a host of other issues.

Eventually it will not be small islands like the Marshall Islands having to choose between relocation, building up the land, or creating high walls to protect it from the ocean, it will be large coastal cities around the world with billions of people affected. Already large cities like New York, Miami, and hundreds of others are spending billions of dollars a year to address damages from rising sea levels and more intensive storm systems.

Ocean CO_2 Levels

The rise in CO_2 and the lowering pH of the oceans are connected. This not only changes the chemistry of the ocean, killing off a host of different life forms, it also creates a feedback loop contributing to Global Warming itself. The more acidic the ocean becomes the less release of dimethyl sulfide (DMS). DMS helps cloud formation, which helps reflect solar radiation.

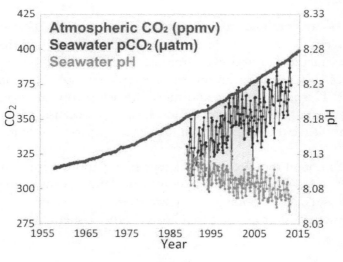

Figure 18

Dying Coral Reefs

Figure 19

Coral reefs cover less than 1% of the ocean floor and yet host nearly ¼ of marine life. They help protect shorelines from waves and tropical storms, provide food and shelter for a great diversity of life, and help fix carbon and nitrogen.

Contributing factors
- warming water causes the polyps to expel the algae living within their tissues
- pollution
- over fishing
- ocean acidification
- chemicals from sunscreens and other products

Marine Life Extinctions

While species die off all the time, the level and speed at which we're seeing living forms become extinct today is troubling.

A case in point is our Chinook Salmon in the Pacific Northwest. The following chart is from a longitudinal study of Chinook populations by the EPA. As can be seen over the 26 years, the Chinook population has been falling drastically.

This is happening to many species in the world. While some species can quickly adapt and benefit from the changing climate, a vast majority are unable to adapt and so are declining in numbers. https://www.iucn.org/resources/conservation-tools/iucn-red-list-threatened-species.

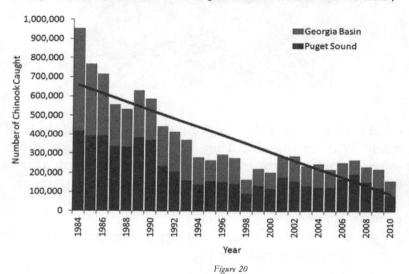

Figure 20

Some factors behind the decline
- warmer oceans are shifting migrational patterns
- warmer stream waters are killing off young salmon and increasing disease rates and parasites
- increasing toxic algae blooms
- floods flushing eggs and young fish from their spawning nests (redds)
- less food due to ocean acidification
- changing seasonal streamflow patterns
- pollutants
- over harvesting
- spawning habitat loss

Oxygen Dead Zones

Figure 21: Gulf of Mexico

Contributing factors
- agricultural, livestock, and wastewater runoff
- with more storms and floods comes more runoff
- toxic algae blooms benefit from nutrient runoff
- rising ocean temperatures
- stagnant waters

Declining Phytoplankton Numbers

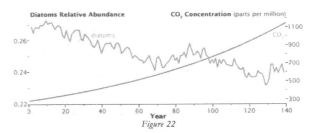

Figure 22

Phytoplankton are single-celled organisms that produce nearly half of our atmosphere's oxygen and are an essential part of carbon sequestering. On top of that, they are a foundational food source for aquatic life as can be seen on the next page.

It is fair to say that much of life as we know it would not exist without phytoplankton. Sadly, like all the other trends we have noticed throughout this guide, phytoplankton numbers are declining. Two major factors in this decline are warming waters and less access to nutrients.

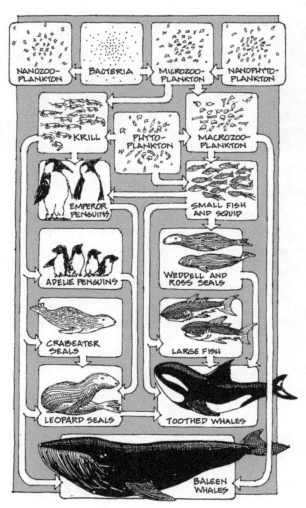

Figure 23

Section IV
FIRE & ICE

Record-breaking fires are becoming common around the world. In the forty years between 1960 and 2000 only two years burned more than six million acres, while ten of the last eighteen years have surpassed that number. On top of that, nine of the top ten worst wildfires in U.S. history have all happened since 2003.

Fires

Changing Forest Fires in the U.S.

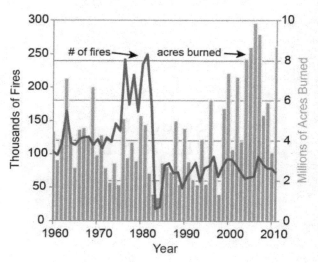

Figure 24

Fires impact Global Warming in many ways. When they happen the carbon stores in the trees are released into the atmosphere as black and brown carbon particulates, there's less trees to capture CO_2, smoke from the fires absorb more solar radiation, while the soot and ash blacken the sea ice, which in turn lessens the amount of reflective surfaces of the earth as well as speeds up their melting due to further absorption of the solar energy.

Greenland Ice Mass

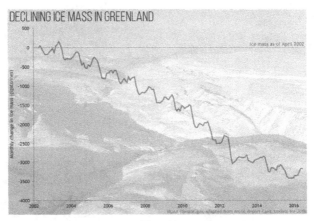

Figure 25

This chart from NOAA shows both the rising and falling mass of ice through the months as can be seen from the step like pattern we have between the years. When looked at as a whole it is easy to see the consistent decline of ice mass through the 14-year period.

Ice mass is measured by satellite imagery and radar (InSar).

The loss of ice mass affects sea levels, water currents, weather patterns, releases methane and other greenhouse gases, and reduces the reflective surfaces of the planet, which in turn reduces the amount of sunlight reflected back into space.

Glacial Retreat

Figure 26

This is one example out of thousands. Over 90% of the world's glaciers are shrinking (https://wgms.ch/global-glacier-state/). As glaciers melt they can create unstable lakes that burst once their icy banks break. This causes massive torrents of ice, rocks, and water to tear down the valleys, ruining crops and killing unsuspecting people below.

Melting glaciers also contribute to rising sea levels.

Millions of people will see their water supplies diminish from melting glaciers even though their personal contribution to Global Warming is miniscule.

Decreasing Snowpack

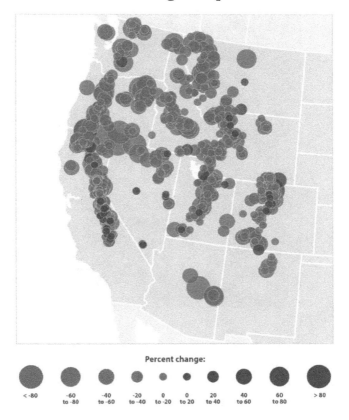

Figure 27

Why is snowpack important?

A large portion of the Western United States and over 50% of the world's population depend on the slow release of water that comes with spring and summer melting. In addition, it helps protect and keep the soil moist throughout the spring and into the summer. This in turn helps minimize the impact of fire.

Figure 28

This decline of snowmelt drastically reduces water levels in rivers and reservoirs. An example of this decline can be seen in the above two satellite images of Lake Powell. The one above was taken in May 2002, and the one below in May 2019.

Section V
MISCONCEPTIONS & MISINFORMATION

Global Warming is a natural phenomenon.

That is true. The difference is thousands of years, or in some cases, mass extinction events. What normally happens over a vast amount of time is happening in just a hundred years or so.

Solar storms are warming things up, we are getting closer to the sun, or our tilt has changed.

Solar storms happen, only, most of the storm's energy is re-emitted back into space. The little bit of energy reaching the lower atmosphere wouldn't even be felt.

Instead of getting closer to the sun, we are slowly pulling away as the sun burns up its mass.

The tilt is definitely changing, as it always has. We just get back to the fact that this natural process happens over thousands of years and not just in a few hundred.

CO_2 has nothing to do with Global Warming.

Take two glasses, pump one with CO_2, place both cups in the sun, and then measure yourself. There is a direct connection.

There is no consensus.

Over 97% of scientists agree Global Warming is happening and that humans have a part to play in it.

Water vapor is what's really heating things up.

That is definitely part of it. The fact that more water vapor arises when things warm up just makes things even worse.

This year was cold: Global Warming isn't real.

Weather patterns are complex and fluctuate from year-to-year and from place-to-place. Take a step back and look at the data. There's no denying that things are getting warmer as a whole.

Human CO_2 emissions are insignificant compared to the emissions coming from nature.

True, but that's only half the picture. Nature is also absorbing that CO_2 through plants, trees, oceans, and such, to the point where it is relatively balanced. What's making things unbalanced is our little bit extra.

It is not economically feasible. It is impossible to get away from fossil fuels without society collapsing.

The cost of not doing something is going to be far worse.

Initially there will be a cost as we transition towards healthier and more sustainable energy sources. Once the infrastructure and everything needed is set up, we will save a lot: monetarily and environmentally.

Even if the whole world did this _____ it would only help 5% or 3% or .3%.

What's the point of doing this _____ if it only helps 3%?

That seems to be the logic. Only, if we did this thing that helps .1%, and this that helps 5%, and this other thing that helps 2%, and so on, at some point things will start to add up and improve.

Nothing is too small. Just ask the camel right before that last piece of straw was placed on its back.

Global Warming is a good thing: longer growing season, opens up shipping lanes, plants are greener, less people will die from the cold, there's more precipitation, and so on.

On any given year we can see someone benefiting from Global Warming. The Russians are currently benefiting with northern shipping lanes opening up as well as more land mass to farm and natural resources to access.

It's hard to argue against these benefits except to say they come with massive floods, increases in devastating fires and extreme weather events, deadly heatwaves, droughts, unpredictable weather patterns that make it harder to guarantee good crop outcomes, and a host of other issues.

While less people die from the cold, we're seeing more people die from heat waves and droughts.

As for the plants getting greener, that is true. We are putting more CO_2 and N_2O into the atmosphere and they love it. Too bad this is not enough to offset all the unwanted stuff going on.

Other things we are noticing are greater contamination and overuse of fresh water supplies, declining global phytoplankton, major changes of migration patterns, disruptions to food chains, forest dieback, and increases in allergens, noxious plants, plant diseases, and insect infestations.

More precipitation comes with the loss of ice sheets and glacial masses, further warming and acidification of the oceans, mudslides, and major flooding events.

A few gains do not outweigh the losses we're experiencing.

Consumers have to change their diets and driving habits.

While this is true, the most impactful change needs to happen at the international level through governments and the industrial complex. This argument is a distraction and another stall tactic.

Scientists are faking the data to keep their jobs.

Even unpaid scientists are saying the same thing.

Alternative energy companies are paying the scientists.

Even some scientists working in the fossil fuel industry acknowledge that Global Warming is a real threat and that humans are a contributing factor

Besides, scientists have been talking about Global Warming way before solar panels.

Other planets are warming: it must be cosmic radiation.

Not all the planets are warming.

Cosmic radiation in space shows regular fluctuating patterns inconsistent with Global Warming. Even when we record massive gamma radiation bursts we do not see a corresponding temperature change. On the other hand, there is some evidence that cosmic radiation contributes to cloud formation, which generally has a cooling effect.

There's nothing that can be done about it, so we might as well enjoy ourselves. Worse still, "I don't care."

There is a lot we can do to address Climate Change while still enjoying life. For those that don't care, I'm sorry to hear that. I hope you find something to care about in life.

The Facts

Things are messed up and they're not getting better. To sum up what we know: global temperatures are rising; the oceans are getting hotter, a growing percentage are suffocating from lack of oxygen, currents are slowing down and changing, the water is becoming more acidic, and sea levels are rising due to thermal expansion, melting glaciers, thinning ice sheets, and coastal erosion; 100 year floods are becoming a yearly occurrence; hurricanes and tornadoes are getting more intense; animal extinctions are happening at a horrifying rate; forests are dying; rivers and reservoirs are shrinking due to lower snowpack levels; fires and droughts are becoming more devastating; old diseases are appearing again and other diseases are spreading as temperatures warm; climate-triggered wars and mass human migrations are becoming more common; the monetary costs are becoming untenable and could lead to economic collapse; and human deaths due to severe weather events are still happening at an alarming rate.

We can push this problem down the road or we can start to address it. It is not going to be easy. Anyone that says otherwise is lying.

Each of us needs to do our part: from world governments to each individual human. The rich countries have a responsibility to take the lead in changing things and in supporting developing countries in creating clean and sustainable infrastructure.

Section VI
THINGS WE CAN DO

First and foremost, educate yourself. The more you know the better off you will be in making informed decisions. After that, start doing something about it.

Climate Change won't be fixed just by shifting our diets or changing all our lights to LED. Climate Change is a complex problem that is going to take a multi-leveled approach to address.

Things need to be addressed from the consumer side of course, but more importantly we need to address the problem on a global scale, and that is only going to be possible with international cooperation and a major shift in industrial thinking. If the world governments do not step up to the plate and enact meaningful measures to stem the tide, no amount of turning off the lights is going to save things.

At the same time it is important to avoid the pitfall of not doing anything because of the false belief that it will not make a difference. Even a 1% difference is a difference. The more little things we do, the closer we will get to making a big difference.

We need to take steps as consumers right now. If the industries will not change, then we need to take our almighty dollar and vote with it by only supporting those companies that actually care about the environment. It might be inconvenient, but the alternative is to continue supporting companies that pollute the world. At that point we are as much to blame as they are.

One thing I have learned by studying Global Warming is that our greatest enemy is inaction. The vastness of the problem can be debilitating. Connect with other concerned people and organizations. Start doing what you can. Talk about it with the people you meet. The more people thinking and talking about Climate Change, the more likely things will change.

Humans are powerful and amazing. We can address Climate Change if we set our minds to it. I really believe in us.

It's not too late as some claim. At the same time it is not so early as to dismiss the signs: things are getting worse every year. If we do not act, and act decisively, there will be a time not so far in the future when we arrive at the tipping point where no amount of effort will stop the cascading effects of Climate Change.

It is better to change now when we have choices and can do it in a way that does not disrupt our lives, than be forced to do it later because we have to and with far less options.

In the next section I share part of my family's ten-year plan towards becoming more responsible citizens of the planet. I do not believe we have to drop everything that we are doing in modern society and go live in caves. I like technology and all that modern society has to offer. I believe we can make meaningful changes without having to negatively impact our quality of life.

We can address Climate Change on five major levels: individually, locally/regionally, industrially (general and agricultural), federally, and internationally. To make an impact on the climate we need to address each of these levels seriously.

The following lists can be daunting and overwhelming. Just start with something, anything. Eventually you will find yourself doing an untold number of things that have a positive influence on the world.

The individual list has check boxes. Occasionally look at the list and see if you can check something off.

Individually

Making home more efficient
- ☐ change to LED bulbs
- ☐ insulate house better
- ☐ buy energy efficient windows and doors
- ☐ turn off electrical devices when not in use
- ☐ turn off lights when not in room
- ☐ use natural light whenever possible
- ☐ regulate house temperature – avoid heating and cooling rooms not being actively used and use a programmable thermostat
- ☐ use fans and open windows instead of using air conditioners when possible
- ☐ close blinds and curtains on sunny side of the house when it is hot to help cool things down
- ☐ invest in alternative energy sources such as wind, solar, and hydropower
- ☐ next time you buy appliances make sure they are energy efficient
- ☐ reseal doors and windows when needed
- ☐ install low-flow showerheads or advanced shower heads like the Nebia
- ☐ tune HVAC system regularly
- ☐ clean or replace house filters
- ☐ use laptops instead of desktops when you can
- ☐ buy low-flush toilets
- ☐ buy on-demand water heater
- ☐ set water heater to 120 degrees
- ☐ solar screens for summer: cover windows in winter
- ☐ live in smaller homes
- ☐ avoid carpets that release VOCs
- ☐ repair leaks and dripping faucets
- ☐ have a green roof

Cleaning
- [ ] run full loads of laundry and dishes
- [ ] wash clothes in cold water
- [ ] hang dry clothes when you can and use air dry setting on dryer when you cannot
- [ ] take quicker showers – avoid excessive running water
- [ ] avoid pre-rising dishes if you have a good dishwasher, if not, then fill sink with water and scrub but don't rinse before placing in dishwasher
- [ ] avoid cleaning products with phosphates, chlorine, and petroleum byproducts (use natural products)
- [ ] clean with rags instead of disposables
- [ ] use recycled unbleached toilet paper and paper products
- [ ] use less soap, toothpaste, detergent, et cetera

Land management
- [ ] lessen sod size to minimize watering needs – or replace with non-invasive drought tolerant ground cover
- [ ] plant trees for shade, food, and carbon sinking
- [ ] plant legumes and other nitrogen fixing plants
- [ ] have a garden
- [ ] compost yard waste instead of burning
- [ ] If you choose to burn brush and stuff, dig hole, place pile in hole, light and let burn for a few minutes and then bury. This creates biochar that is good for soil and plants and helps with carbon sequestering
- [ ] don't use pesticides, herbicides, or other nasty chemicals
- [ ] use natural fertilizers, not synthetic ones
- [ ] plant drought-tolerant plants and trees
- [ ] mulch
- [ ] Permaculture
- [ ] Hugelkulture

- ☐ aerate areas that are waterlogged to help minimize methane release and nitrogen denitrification
- ☐ minimize power tool use and when needed move towards electric tools
- ☐ grow ground cover
- ☐ plant things that encourage animals and beneficial insects
- ☐ plant more trees, lots and lots of trees
- ☐ if you have forested areas, let them be
- ☐ minimize areas of erosion
- ☐ encourage biodiversity
- ☐ avoid invasive plants
- ☐ use solar-powered lighting
- ☐ reduce water consumption: irrigating, mulching, deep weekly watering to encourage root growth instead of daily shallow watering, timing to minimize evaporation, planting drought tolerant plants, and catching rainwater are all good practices

Food

- ☐ use microwave instead of the stove when possible
- ☐ use the right size cooking element for the pot
- ☐ use a lid while simmering or boiling to reduce the amount of energy needed to cook
- ☐ start composting, create worm bins, or better yet, bury food directly in the garden
- ☐ eat the food you buy instead of letting it go to waste
- ☐ eat raw food more, cook less
- ☐ cook with cast iron
- ☐ strain and reuse fat from meat
- ☐ eat less meat and dairy which does many good things, some of which are reducing methane, using less land for growing animal feed, protecting land and water ways, and minimizing manure management needs

- ☐ cook instead of going out to eat
- ☐ re-use storage containers
- ☐ buy in bulk and support local farms
- ☐ avoid fast foods
- ☐ bake with solar ovens

Transportation
- ☐ walk or bike more
- ☐ use public transportation
- ☐ carpool and do rideshares
- ☐ buy hybrid and electric cars
- ☐ avoid air travel or fly coach
- ☐ drive less
- ☐ do all shopping in one trip
- ☐ vacation closer to home
- ☐ check tire pressure
- ☐ avoid idling
- ☐ open windows instead of using air conditioner
- ☐ keep car tuned
- ☐ change filters
- ☐ check exhaust system for leaks
- ☐ buy locally made products to reduce shipping needs
- ☐ conference calling when possible
- ☐ avoid rush hour and stop-and-go traffic when possible
- ☐ avoid speeding (stay within the 55-60 mph range to optimize fuel efficiency and minimize emissions)
- ☐ take unnecessary weight out of vehicles

Other things
- ☐ educate yourself and share what you learn with others
- ☐ get involved with organizations that are doing something
- ☐ write to your representatives and demand more action

- [ ] buy less things; when you do buy things, buy things that will last and are environmentally friendly
- [ ] reduce, reuse, and recycle
- [ ] avoid excessive packaging
- [ ] if you don't care to collect books and movies, purchase electronic streaming versions instead
- [ ] shop at second hand stores
- [ ] buy shoes and other products made with recycled materials
- [ ] support Earth conscious businesses by buying their products
- [ ] use reusable bags when shopping
- [ ] reduce waste
- [ ] teach children and lead by example
- [ ] download apps that help awareness such as Kil-Ur-Watts, Wider EMS, Greenpeace's Click Clean, et cetera
- [ ] conserve water and energy whenever you can
- [ ] encourage your employer to move towards more energy efficiency
- [ ] take a moment daily to breathe, appreciate, and give thanks for the blessings you have

Regionally and Locally

Political
- vote in politicians that are going to address Climate Change
- hold politicians accountable for misinformation

Land Management
- protecting state lands and resources
- restore key ecosystems
- support small organic agricultural producers
- drought resistant landscaping

- create green roofs, roadside planting, rain gardens, edible parks, et cetera
- encourage community gardens
- protecting floodplains and water ways
- stopping urban sprawl
- creating rain water catchment systems

Energy
- change city lights to LED
- all government buildings have LED lights
- shifting fleet to electric vehicles
- wisely managing building temperatures
- creating own energy
- all new buildings are energy efficient
- requiring utility companies to generate a higher percentage of their electricity through renewable energy sources
- support energy buy-back programs where electric companies can buy energy off of their customers that produce their own with solar panels and so forth
- creating micro-grids

Infrastructure
- create earth-friendly cities and infrastructure
- build with light-colored roofing and pavement to reflect solar radiation
- all new buildings are energy efficient (insulated properly, smart glass windows, wind tower cooling technology, solar mirror technology for lighting during the day, producing own electricity (solar, wind, et cetera), more stairs, radiant floor heating, solar and on-demand water heating, motion sensor facets, no heat hand dryers, and so on
- improve water supply, drainage, and treatment

- infrastructure
- quickly repairing leaks of any sort: water, gas, sewage, et cetera
- city planning for more room to walk and bike
- update infrastructure towards more efficiency
- updating waste management plants
- enacting strong recycling programs
- expanding affordable housing
- strengthening public transportation
- create more green spaces
- permeable pavement
- micro-wind turbines on major freeways to catch wind from passing cars to generate electricity for lights and the power grid

Other
- reducing waste
- collaborating with community groups
- encourage urban density and safety to encourage walking instead of driving
- encourage local industry
- educate constituents
- starting or expanding farmer markets
- build resilience and shore up areas of vulnerability to climate change
- have plans in place and funding for disaster relief
- ban plastic bags
- support family planning programs and improve contraception access
- encourage companies to move towards sustainable business models that are environmentally responsible
- funding and creation of programs that reduce food waste

Industry

Energy
- creating own energy: solar, wind, et cetera
- improving energy efficiency
- less heating and air conditioning use
- updating equipment
- using LED bulbs
- use cloud technologies
- all new buildings are energy efficient (insulated properly, smart glass windows, wind tower cooling technology, solar mirror technology for lighting during the day, producing own electricity (solar, wind, et cetera), more stairs, radiant floor heating, solar and on-demand water heating, motion sensor facets, no heat hand dryers, and so on
- improve data center's operational, hardware, and infrastructure efficiency
- continue to improve energy storage and transportation technologies
- working with concentrated solar power in making steel, concrete, and other high heat products

Waste
- reduce waste and pollution and recycle when possible
- move towards environmentally-sound packaging
- move towards producing products that are renewable and made of recycled materials
- create composting system for employees' food waste
- reduce water waste
- use less paper

Transportation
- buy electric company cars
- encourage car pooling
- shorten work week
- allow for video conferencing and telepresence to help minimize travel
- fly coach
- improving ship efficiency
    - operating at efficient speeds
    - coating technologies to minimize drag and keep hulls free of marine creatures
    - wind propulsion systems
    - better fuels
    - improving propelling systems
    - using exhaust scrubbers
- improving airplane efficiency
    - better fuel efficiency
    - using biofuels
    - more passengers on each plane
    - better propulsion technologies
    - using lighter materials
    - improving operational practices
    - improving aerodynamics
    - more flexible flight plans to take advantage of favorable weather conditions
    - continuous ascents and descents
    - improving airport designs
    - using solar panels and LED lights throughout airport
- improving freight truck efficiency
    - electric trucks outfitted with solar panels on trailers
    - improving aerodynamics
    - improving combustion technologies
    - driving at most efficient speeds
    - improving operational practices

-
  -
    - not idling while sleeping
    - better insolated cabs
    - using alternative fuels
    - better tires
    - tuning and keeping up with maintenance
    - lighter materials
    - shipping more at a time
  - improving train efficiency
    - moving to high-speed rails
    - more efficient fuel-use
    - better aerodynamics
    - hybrid engines
    - improving stop-start technologies
    - managing speed to improve fuel efficiency
    - transporting more per trip
    - improving wheels and rails to minimize friction
    - utilizing software technologies to improve operations

Materials

- Developing climate action plans that address supply chain issues: a recent example is LEGO toymaker shifting from petro-plastics to bio-based materials.
- utilizing recyclable or compostable packaging such as mushroom packaging
- when catering focus more towards vegetarian foods and less on red meats and processed foods
- choose environmentally conscious vendors
- source materials from renewable resources and stop commodity-driven deforestation from all supply chains

Other

- build employee awareness
- empower employees to create innovative ways to address Climate Change
- not putting profits above the Earth's health

- measuring carbon footprint, being transparent, and making an honest effort towards reducing emissions
- implementing and improving carbon capture and storage technologies
- creating sustainable and responsible business models that support smart climate policies
- invest in innovation

Agricultural Industry

Farming Practices
- moving towards smaller, local farms
- moving towards organic practices
- regenerative agriculture
- utilizing cover crops to prevent erosion and to build soil
- utilizing natural fertilizers and moving away from synthetics
- nutrient management (timing fertilization and using the proper amounts to minimize waste and run-off)
- utilizing nitrogen fixing plants to enrich soil
- crop rotation
- companion planting
- building and feeding the soil to increase earthworms and microorganisms (fungi, protozoa, nematodes, beetles, springtails, and some bacteria and actinomycetes are a few examples)
- breeding drought tolerant plants
- reducing rice emission by improving water and nutrient management, as well as, improving rice straw management
- reducing tillage or moving to no-till systems (tilling kills microorganisms that support plant life, causes nutrient loss by wind and water, increases unwanted bacteria and actinomycetes, increases CO_2 and N_2O emissions,

disrupts soil cycles, reduces mycelium growth, and a bunch of other unwanted things)
- silvopasture (raising animals with trees)
- pasture cropping (integrating perennial grains into pasture lands)
- sustainable intensification farming
- efficient water use and finding methods of replenishing ground water
- vertical gardening
- lab grown proteins

Land and Resource Management
- better water management
- capture and storing rainwater
- better manure management programs (digesters programs and aerobic composting of manure are two examples)
- sustainable grazing management
- stop burning down forests
- agroforestry
- sustainable lumber practices
- move away from messing with peatland and replace seedling mixtures with more sustainable coco coir

Livestock
- improving livestock diets and using feed additives to reduce methane
  https://www.agric.wa.gov.au/climate-change/carbon-farming-reducing-methane-emissions-cattle-using-feed-additives
- optimizing animal health
- improve land fertility for grazing
- integrating livestock into crop production

Other stuff
- creating resiliency and planning to address changing climate
- reducing food waste
- increasing biofuel production
- streamlining supply-chain practices
- upgrading industrial equipment

Federally

Politically
- elect leaders that bring awareness and actions to the table
- create laws that punish politicians for spreading misinformation

Policies and Regulation
- push for better fuel standards
- setting higher emission standards
- stronger environmental laws and funding enforcement agencies that hold companies accountable
- enact carbon cap-and-trade laws
- better national forest management laws
- national recycling standards

Research and Education
- mandating school education on global responsibility and citizenship
- address misinformation
- more research into alternative fuels and building materials
- researching better energy storage and transportation technologies

- research better and more efficient recycling technologies
- research geo-engineering and bio-engineering methods
- having banks and insurers do "Stress Tests" to insure they are resilient and able to deal with severe climate events
- researching capture and neutralization technologies
- Researching methods to lower life-cycle costs in production and distribution, and enacting plans to address these issues.
  - *Life-cycle cost*: the environmental cost of making a product, such as a solar panel in some factory in China, and the time it will take to reach a "net-zero" level of emission can be a long time. We have to account for the environmental impact of mining the materials, transporting them between one place and another, transforming them into some specific component and then shipping them again to another place and another place before it all comes together in some factory to be made into a solar panel and then again shipped to another place and then another place, before ultimately being shipped overseas to be held in some warehouse before going to a store where some conscientious person buys it. If you also account for repairs and all that goes into its disposal, reaching "net zero" might very well take some time. It's still better than what we're doing now. Even so, figuring out ways to be more efficient and environmentally responsible needs to be a priority.
- provide assistance and training for workers who transition from high to low carbon producing industries
- periodic scientific reviews of implemented actions and adjusting when needed

Investing
- allocating more money for research, development, and implementation of low-emitting technologies
- massive tax incentives for cleaner and more efficient choices
- more funding for alternative energies (solar, thermal, hydro, geothermal, tidal, wind, gas digesters, nanogenerators, cogenerative systems, et cetera)
- funding for retrofitting buildings
- investing in sustainable public housing
- funding for capture and neutralization technologies
- lessening fossil fuel subsidies
- fund and help prepare communities for the climate changes already on the way
- financial assistance for poor and middle income Americans to transition towards sustainability
- incentives for sustainable agriculture and private forest management
- funding for state and local communities to replace city lights and government buildings lights with LEDs
- investment in high-speed rails

Energy
- shift towards all electric cars for federal fleet
- LED bulbs in all federal buildings
- move towards creating own energy (solar panels, wind, et cetera)
- regulate building temperature
- all new buildings are energy efficient (insulated properly, smart glass windows, wind tower cooling technology, solar mirror technology for lighting during the day, producing own electricity (solar, wind, et cetera), more stairs, radiant floor heating, solar and on-demand water heating, motion sensor facets, no heat

hand dryers, and so on
- building better power grids
- better batteries
- low carbon jet fuel
- phasing out fluorinated gas
- update and continue running current nuclear plants (still environmentally toxic, but at least not as devastating to the atmosphere as fossil fuel burning – and they are already in place)
- making grids smarter to decrease energy loss, making transmissions more efficient, more seamless integration with customer-owned power systems, increasing customer awareness and opportunities to cut power usages, mitigating emissions, saving money, and a host of other great benefits

Internationally

- support international studies and organization such as IPCC, United Nations Convention on Climate Change, and 2019 Climate Action Summit
- encourage international cooperation
- push for more international agreements such as Kyoto Protocol, Paris Agreement, Montreal Protocol, Kigali Accord
- demand that the government become part of the solution by joining other nations on finding ways to address the issues
- pool money from richer nations to help fund developing countries in building earth friendly infrastructure and alternative energy systems such as efforts like "Shifting the Trillions"
- encouraging and supporting all nations to empower and educate women

Our Ten-Year Plan

There are people out there that demand we stop driving, flying, and eating meat. While that would make a huge impact on reducing greenhouse gases, the fact is, that's not going to happen. The only way that kind of change happens is if we're forced to do it. We humans are far too comfortable with our lifestyles to make such a drastic change collectively.

For me personally, I don't want to feel like I am giving up things, to feel like I'm losing out somehow. I want to feel good about my choices and to feel like I am gaining something. Instead of feeling like I'm sacrificing my way of life or giving up things that I like, I choose to trade up, which is fulfilling.

One issue I have with the Climate Change conversation is that we're focused on the wrong thing. We need to change the narrative. Instead of doing things because we have to, we do them because we want to, because it's the right thing to do, and because it feels good.

So we shift the focus. Instead of our choices arising out of fear, they come from joy and a strong desire to take care of our inheritance: the Earth.

My family is on a ten-year plan. In that time we aim to surpass net-zero targets by reducing our emissions and helping to heal our Planet. The first step we took on this journey was the easiest: we changed over to LED bulbs. This saved us money and reduced the amount of energy we use.

Since I started studying Climate Change I've also started turning off lights and electrical devices when they're not needed. This includes such things as shutting the fridge instead of keeping it open for a second to do something, and turning off the water while brushing my teeth.

When it comes to transportation we were able to get a hybrid car to help reduce the amount of gas we needed to buy, as well as reduce our emissions with smart car technology. Whenever we go out, we make an effort to do all of our shopping in one trip. This saves us time and money, and reduces our emissions.

It's cold in the winter. We warm ourselves with electric heaters and a wood stove. We only heat the living room during the day to minimize the amount of energy we use. This upcoming year we will focus on insulating our house better and in the future purchase an indoor mass rocket stove that will optimize heating, reduce our need for fuel, while at the same time nearly eliminating CO_2 emissions.

Food production, distribution, and waste are huge contributors to greenhouse gas emissions. In this area of our lives we're starting to buy bulk groceries with minimal packaging, as well as focusing on shifting our diet more towards whole grains and veggies. We compost all food remnants for our garden, and every year we strive to grow a little bit more until we are able to grow, catch, hunt, and raise most of the food we eat.

While we have been recycling for years and trying to minimize our trash, this upcoming year we will make an effort to reduce our waste even more. Our family of five still throws away about 55 gallons worth of trash every two weeks. Next year our goal is to reduce that to once a month.

Our next step towards saving money and reducing our energy emissions is to buy energy efficient appliances, tools, and electronics. A few things on our list are an on-demand water heater, battery operated garden tools, all-weather panels (energy from sun and rain), and electric vehicles.

I personally don't want to clear a fallen tree with an axe made from a rock, stick, and cordage just to zero my emissions. I could do it, but I'd rather have an electric chainsaw.

These are some of the things we are doing to address Climate Change. Every year we continue to check off things from the list starting on page 56. Within ten years it is our goal to do everything we can to ensure our great-great-grandchildren will still have a healthy planet to call home.

These little changes might seem small, but like drops of water filling a bucket, every little drop adds up. The more people filling buckets, the quicker we can cool things down.

SOURCES FOR GRAPHS

I want to thank all the scientists, companies, government employees, and all those that have made an effort to educate us about Climate Change. You have given us the data and tools needed to make a positive change.

I take responsibility for my part and promise to make an effort to do what I can to make things better.

1. https://earthobservatory.nasa.gov/features/GlobalWarming/page2.php
2. https://www.climate.gov/news-features/understanding-climate/climate-change-atmospheric-carbon-dioxide
3. https://upload.wikimedia.org/wikipedia/commons/a/a3/Mlo_ch4_ts_obs_03437.png
4. https://www.epa.gov/sites/production/files/styles/medium/public/2019-04/gases-by-ch4-2019.jpg
5. https://www.eia.gov/environment/emissions/ghg_report/images/figure22-lg.jpg
6. https://www.epa.gov/ghgemissions/overview-greenhouse-gases#f-gases
7. https://commons.wikimedia.org/wiki/File:BAMS_climate_assess_boulder_water_vapor_2002.png
8. https://www.climate.gov/news-features/blogs/beyond-data/2018s-billion-dollar-disasters-context
9. https://www.epa.gov/climate-indicators/climate-change-indicators-coastal-flooding
10. https://www.nrcs.usda.gov/wps/portal/nrcs/detail/national/nedc/training/soil/?cid=nrcs142p2_054003
11. https://commons.wikimedia.org/wiki/File:Desertification_Control_Project,_Ningxia_China_-_Planet_Labs_satellite_image.jpg
12. https://www.ncdc.noaa.gov/sotc/tornadoes/201613

13. https://www.epa.gov/climate-indicators/climate-change-indicators-drought
14. https://www.nasa.gov/feature/ames/aerial-images-show-decades-of-foothill-forest-growth-erased-due-to-california-s-extreme
15. https://www.nhc.noaa.gov/climo/
16. https://www.epa.gov/climate-indicators/climate-change-indicators-sea-surface-temperature
17. https://www.epa.gov/climate-indicators/climate-change-indicators-sea-level
18. https://oceanacidification.noaa.gov/OurChangingOcean.aspx
19. https://commons.wikimedia.org/wiki/File:Bleached_coral_(24577819724).jpg
20. https://www.epa.gov/salish-sea/chinook-salmon
21. https://commons.wikimedia.org/wiki/File:Dead-Zone_-_Gulf-of-Mexico.jpg
22. https://earthobservatory.nasa.gov/ContentFeature/Phytoplankton/images/diatoms_CO2.png
23. https://earthobservatory.nasa.gov/features/Phytoplankton
24. https://www.globalchange.gov/browse/multimedia/changing-forest-fires-us
25. https://www.noaa.gov/sites/default/files/thumbnails/image/INFOGRAPHIC-greenland-decline-12.12.16-climate.gov-%20700x502%20-%20inset.png
26. https://www.climate.gov/sites/default/files/CareserGlacier_1933-2012_610.jpg
27. https://www.epa.gov/climate-indicators/climate-change-indicators-snowpack
28. https://earthobservatory.nasa.gov/world-of-change/LakePowell

Cover Image

https://eoimages.gsfc.nasa.gov/images/imagerecords/0/885/modis_wonderglobe_lrg.jpg

RESOURCES

IPCC Climate Change Report
https://www.ipcc.ch/sr15/

IPCC Historical Overview of Climate Change Science
https://www.ipcc.ch/site/assets/uploads/2018/03/ar4-wg1-chapter1.pdf

Books

Drawdown: The Most Comprehensive Plan Ever Proposed to Reverse Global Warming edited by Paul Hawken
The Uninhabitable Earth by David Wallace-Wells
The Sixth Extinction by Elizabeth Kolbert
The Soil will Save Us by Kristin Ohlson
Building a Better World in Your Backyard by Paul Wheaton

Documentaries

An Inconvenient Truth
Before the Flood
Chasing Ice
11th Hour
A Beautiful Planet

Good Internet Sources

NASA Global Climate Change
https://climate.nasa.gov/

Skeptical Science
https://skepticalscience.com/

IPCC
https://www.ipcc.ch/

Yale Institute Research on Climate Change
https://climatecommunication.yale.edu/publications/?_sft_ra-format=report

EPA Climate Change
https://www.epa.gov/sites/production/files/signpost/cc.html

UN Climate Action
https://www.un.org/en/climatechange/

NOAA
https://www.ncdc.noaa.gov/

Center for Climate and Energy Solutions
https://www.c2es.org/

Podcasts

America Adapts
Climate Cast
Climate One
Mother of Invention
Reversing Climate Change
Warm Regards

Some Eco-Friendly Companies

Bags and purses
- CGC
- Cuyana
- EST WST
- Everlane
- Fjallraven

- grünBAG
- Mariclaro
- O My Bag
- Sandqvist
- Solgaard
- Terra Thread
- thredUP
- United By Blue

Clothing
- Second hand stores
- Aeon Row
- Alternative Apparel
- Amour Vert
- Boden
- Cienne
- DL 1961
- Ecovibe
- Fair Trade Winds
- G Star Raw
- LA Relaxed
- Ninety Percent
- Organic Basics
- Outerknown
- Outdoor Voices
- Pact
- People Tree
- prAna
- Reformation
- Tentree
- Thought Clothing
- Threads 4 Thought
- United By Blue

Shoes
- Allbirds
- Everlane
- Native Shoes
- Nisolo
- TOMS
- Veja

Skin care and Makeup
- 100% Pure
- Alima Pure
- Axiology
- Antonym Cosmetics
- Besame
- Elate Cosmetics
- Juice Beauty
- Kjaer Weis
- Tata Harper
- W3LL People

Paper and Cleaning Products
- Common Good
- Ecos
- Method
- Seventh Generation

Bathroom Products
- Albatross razors
- Hydrophil
- Georganics
- Lucky Teeth floss
- Moso bamboo

- Naked Necessities
- No. 2
- Seventh Generation
- Simple Soney
- The Humble Co.
- Tom's
- Unpaste

Kitchen Products
- Bee's Wrap
- Biobag
- ChicoBag
- Full Circle Home
- If You Care
- Silicone mats/lids instead of foil for baking
- Stasher

Furniture and Home Goods
- ABC Home
- Avocado
- Burrow
- Crate & Barrel
- Etsy Reclaimed Furniture
- IKEA
- Inmod
- Joybird
- Medley
- VivaTerra
- West Elm

Bedding
- Amerisleep
- Avocado Green

- Brentwood Home
- Happsy
- Naturepedic
- My Green Mattress
- Saatva

Some Eco-product Websites
- https://theuprootedrose.com/ecoliving
- https://earthhero.com/
- https://eartheasy.com/
- https://lifewithoutplastic.com/
- https://greenheartshop.org/

This is a partial list of the many companies with eco-friendly business models. The best way to find what you want is to search the internet. Type in what you want and the words eco-friendly: for instance, eco-friendly earphones.

Some things to consider when supporting a company
- fair trade
- fair wages
- biodegradable packaging
- zero packaging
- renewable energy
- organic materials
- sustainable materials
- recycled materials
- upcycling
- giving back
- water conservation
- zero-emission
- zero waste

ABOUT THE AUTHOR

Tobias lives in the Evergreen State with his wife and amazing children. He loves gardening, playing in the woods, swimming in rivers, and hanging out with family and friends.

Tobias homeschools his two youngest children. This year's science focuses on Meteorology and Climatology.